OXFORD LOGIC GUIDES: 42

General Editors

DOV M. GABBAY
ANGUS MACINTYRE
DANA SCOTT

OXFORD LOGIC GUIDES

1. Jane Bridge: *Beginning model theory: the completeness theorem and some consequences*
2. Michael Dummett: *Elements of intuitionism* (*1st edition*)
3. A.S. Troelstra: *Choice sequences: a chapter of intuitionistic mathematics*
4. J.L. Bell: *Boolean-valued models and independence proofs in set theory* (*1st edition*)
5. Krister Seberberg: *Classical propositional operators: an exercise in the foundation of logic*
6. G.C. Smith: *The Boole–De Morgan correspondence 1842–1864*
7. Alec Fisher: *Formal number theory and computability: a work book*
8. Anand Pillay: *An introduction to stability theory*
9. H.E. Rose: *Subrecursion: functions and hierarchies*
10. Michael Hallett: *Cantorian set theory and limitation of size*
11. R. Mansfield and G. Weitkamp: *Recursive aspects of descriptive set theory*
12. J.L. Bell: *Boolean-valued models and independence proofs in set theory* (*2nd edition*)
13. Melvin Fitting: *Computability theory: semantics and logic programming*
14. J.L. Bell: *Toposes and local set theories: an introduction*
15. R. Kaye: *Models of Peano arithmetic*
16. J. Chapman and F. Rowbottom: *Relative category theory and geometric morphisms: a logical approach*
17. Stewart Shapiro: *Foundations without foundationalism*
18. John P. Cleave: *A study of logics*
19. R.M. Smullyan: *Gödel's incompleteness theorems*
20. T.E. Forster: *Set theory with a universal set: exploring an untyped universe* (*1st edition*)
21. C. McLarty: *Elementary categories, elementary toposes*
22. R.M. Smullyan: *Recursion theory for metamathematics*
23. Peter Clote and Jan Krajíček: *Arithmetic, proof theory, and computational complexity*
24. A. Tarski: *Introduction to logic and to the methodology of deductive sciences*
25. G. Malinowski: *Many valued logics*
26. Alexandre Borovik and Ali Nesin: *Groups of finite Morley rank*
27. R.M. Smullyan: *Diagonalization and self-reference*
28. Dov M. Gabbay, Ian Hodkinson, and Mark Reynolds: *Temporal logic: mathematical foundations and computational aspects: Volume 1*
29. Saharon Shelah: *Cardinal arithmetic*
30. Erik Sandewall: *Features and fluents: Volume I: a systematic approach to the representation of knowledge about dynamical systems*
31. T.E. Forster: *Set theory with a universal set: exploring an untyped universe* (*2nd edition*)
32. Anand Pillay: *Geometric stability theory*
33. Dov. M. Gabbay: *Labelled deductive systems*
34. Raymond M. Smullyan and Melvin Fitting: *Set theory and the continuum problem*
35. Alexander Chagrov and Michael Zakharyaschev: *Modal logic*
36. G. Sambin and J. Smith: *Twenty-five years of Martin-Löf constructive type theory*
37. María Manzano: *Model theory*
38. Dov M. Gabbay: *Fibring logics*
39. Michael Dummett: *Elements of intuitionism* (*2nd edition*)
40. D.M. Gabbay, M.A. Reynolds, and M. Finger: *Temporal logic: mathematical foundations and computational aspects* volume 2
41. J.M. Dunn and G. Hardegree: *Algebraic methods in philosophical logic*
42. H. Rott: *Change, Choice and Inference: a study of belief revision and nonmonotonic reasoning*